ESPIONAGE BLACK BOOK EIGHT

In this series of technical monographs:

ESPIONAGE BLACK BOOK EIGHT:

Industrial Espionage Explained

Expanded Edition

Henry W. Prunckun

Bibliologica Press

Espionage Black Book Eight:
Industrial Espionage Explained—Expanded Edition
by Henry W. Prunckun

ISBN 978-0-6456209-7-9

A catalogue record for this
book is available from the
National Library of Australia

Bibliologica Press
P.O. Box 656
Unley, South Australia, 5061
Australia

CONTENTS

— CHAPTER ONE —

Industrial Espionage

A spy-for-hire, working on a contract for a boutique brewing company, successfully obtained the trade secrets of her employer's competitor. She used charm and cunning to gain the trust of a mid-level manager at the competitor's firm, who granted her access to sensitive information regarding the company's secret beer recipe.[1]

Using her cell phone's camera, the spy made copies of the recipe, process, and ingredients and passed them on to her employer, who quickly adapted it to produce a similar beer.[2]

The competitor realized their recipe had been compromised only after the new brew's launch. Although they started an investigation, they would never determine the source of the leak.

In the following months, the boutique brewer saw a significant increase in sales, establishing it as a major player in the industry. The spy's actions proved to be highly profitable.

Definition and Purpose

Industrial espionage is stealing sensitive information from a business to gain a competitive advantage. This information can range from manufacturing processes and

---

1. Although notional, this example is based on real-life events.

2. As storage media improves, it allows more data to be held on devices. This, in turn, makes it easier to steal more information.

formulas to customer lists and marketing strategies. Industrial espionage is conducted by companies competing in the same market or by foreign governments. Industrial espionage aims to acquire information that will provide an advantage to the business that seeks it at the expense of the company whose information was stolen.

The motivations behind industrial espionage can vary, ranging from the desire to increase profits to the need to acquire critical technology for military or national security purposes.[3]

A trade secret is a piece of information treated as confidential by an enterprise because its particular features combined with limited access provide a competitive advantage. Such a secret piece of information can be durable or ephemeral, so long as it helps enterprises to perform better, faster or at lower cost.[4]

An industrial spy is used because it would be too expensive or technically impossible to obtain through legitimate means (e.g., to develop the information themselves). This information can then be used to improve existing products, create new products, or enter new markets. In some cases, industrial espionage can also

---

3. Carl Roper, *Trade Secret Theft, Industrial Espionage, and the China Threat* (Boca Raton: CRC Press, 2014).

4. International Chamber of Commerce, *Protecting Trade Secrets—Recent EU and US Reforms* (Paris: ICC, 2019), p. 7.

be used to sabotage the competition by spreading false information or disrupting their operations to get ahead.[5]

While industrial espionage can provide a competitive advantage, it is illegal in many countries and can result in legal and financial consequences for those caught engaging in this questionable activity.

Historical Overview

Industrial espionage, also known as corporate spying, has a long history that dates to the early days of trade and commerce. The origins of industrial espionage can be traced back to the ancient world, where merchants gathered information about their competitors. This practice continued throughout the Middle Ages, with the development of trade guilds, which often engaged in espionage to protect their interests. With the advent of the industrial revolution in the eighteenth and nineteenth centuries, the scale and scope of industrial espionage increased in response to the race for technological advancement and economic dominance stepped-up.

During World War I, industrial espionage became more sophisticated, with governments and corporations using spies to gather information about their rivals. The use of intelligence agencies to collect economic and industrial information became more widespread during the Second World War. It continued in the post-war period as the Cold War intensified. During this latter period, the focus of industrial espionage shifted from trade secrets to technology because the need for technological superiority

---

5. This is analogous to a runner tripping a competitor to win the race.

between the East and West became a critical factor in global affairs.[6]

With the rise of the digital age, the landscape of industrial espionage changed once again. The widespread use of computers, the Internet, and other digital technologies have made illegally obtaining and transferring sensitive information easier than ever.

Globalization has made it easier for companies to operate in many countries, increasing cross-border industrial espionage (i.e., a form of transnational crime).[7] The focus of industrial espionage then shifted from state-sponsored activities to criminal organizations, which use hacking and other forms of penetration to obtain trade secrets. As such, the threat of industrial espionage remains a significant concern for companies as well as governments alike. This makes protecting proprietary information and intellectual property (IP) a business imperative.

Why Industrial Espionage

It is easy to see from the notional case of the theft of a beer recipe at the start of this chapter that industrial espionage can impact those who use it and those subjected to it. These impacts cannot be overstated—they directly impact a company's competitiveness and reputation.

For those targeted by industrial spying, losing sensitive data will affect their financial "bottom line." These

---

6. The Former-Soviet Union's launch of the Sputnik satellite could be seen as a symbol of this competition.

7. Paola Bertucci, "Enlightened Secrets: Silk, Intelligent Travel, and Industrial Espionage in Eighteenth-Century France," in *Technology and Culture*, Issue 54, Number. 4 (2013): pp. 820–852.

impacts can harm a company's relationship with its customers, investors, and partners, reducing trust in its products and services.

For those that obtain classified data about a competitor's products or services, prices, marketing strategies, or other trade secrets, it can help that company make better business decisions, such as adjusting its prices, formulating new marketing strategies, or changing its product offerings to improve its position in the marketplace.[8] A company can save vast amounts of money and perhaps years of development work by stealing a competitor's information. This is particularly true in fields like aerospace and pharmaceuticals.

In addition to the negative impacts on a company, industrial espionage can have broader implications for national security and the global economy. For example, the harvest of intellectual property by state-sponsored industrial espionage can be passed onto that country's businesses or military industries for strategic purposes. These acts can create imbalances in the global marketplace and undermine the stability of the international economy.

However, if companies are aware of the risks of industrial espionage, they can take steps to protect their confidential information.[9] We will discuss security in a later chapter. Still, to provide an idea of the gambit, they include data encryption, computer and network access

---

8. William C. Hannas, James C. Mulvenon, and Anna B. Puglisi, *Chinese Industrial Espionage: Technology Acquisition and Military Modernization* (Abingdon, Oxon: Routledge, 2013).

9. Henry W. Prunckun, *Information Security: A Practical Handbook on Business Counterintelligence* (South Australia: Bibliologica Press, 2019).

controls, audits, and educating employees on the dangers of industrial espionage and the signs of attempts at penetration.

Although cases of industrial espionage appear in the media, the phenomenon is not recent. Nevertheless, the rise of industrialization followed by globalization has increased the economic stakes—small and large businesses now compete globally for market share. A few examples will show the unsparing nature of business competition and the lengths companies go to secure an advantage over their rivals.

One of the earliest recorded instances of industrial espionage was in ancient Greece, where the King of Syracuse commissioned the philosopher and mathematician Archimedes to develop weapons for a potential war with Rome. According to legend, the Roman general Marcus Claudius Marcellus ordered spies to steal the designs, leading to the eventual capture of Syracuse and the death of Archimedes.[10]

In the early modern era, the Dutch East India Company was one of the world's largest and most powerful trading companies. The company used spies to gather information on its competitors and steal trade secrets to maintain its monopoly. The company's spies were so successful that they even obtained the secret process for Chinese porcelain, which they then manufactured in the Netherlands.[11]

---

10. André Gerolymatos, *Espionage and Treason in Classical Greece: Ancient Spies and Lies* (Lanham, MD: Lexington Books, 2019).

11. T. Volker, *Porcelain and the Dutch East Inia Company* (Leiden, Holland: E.J. Brill, 1971).

Another case involved Robert Fortune, a nineteenth-century botanist. According to Rose, he perpetrated the "...greatest theft of trade secrets in the history of mankind"—the theft of tea plants and seeds from China. Disguising himself in Mandarin robes, he ventured into a tea-growing region of China to commit industrial espionage. His haul of botanical specimens and the knowledge to grow them was then taken to India, part of the British Empire.[12]

The Twentieth Century saw the rise of government-sponsored industrial espionage as countries sought to gain an advantage in the global marketplace. During World War II, the Allies and the Axis powers used spies to gather information on each other's military-industrial capabilities. In particular, the Allies used code-breaking to intercept and decode encrypted German messages,[13] while the Germans used spies to infiltrate British military industrial facilities.

Industrial espionage was a concern for businesses and governments in the post-war era. In the 1970s and 1980s, the former Soviet Union was one of the largest and most active practitioners of industrial espionage, using spies to steal technological and industrial secrets from Western countries. One of the most famous cases was the "Illegals Program," in which Soviet spies posing as ordinary

---

12. Sarah Rose, *For All the Tea in China: How England Stole the World's Favorite Drink and Changed History* (New York: Viking, 2010), p. 272.

13. David Kahn, *The Code-Breakers* (Toronto: The Macmillan Company, 1967).

citizens infiltrated businesses and government organizations in the West.[14]

The rise of globalization, which brought interconnectedness to the world's economies, has led to a new era of industrial espionage. In 2010, the U.S. Department of Justice charged several Chinese telecommunications company Huawei employees with stealing trade secrets from U.S. companies. In 2014, the German car manufacturer Volkswagen was accused of cheating on emissions tests, leading to a scandal that cost the company billions of dollars and damaged its reputation.

Industrial Espionage versus Business Intelligence

Industrial espionage and business intelligence are two types of inquiries used to gather information for commercial purposes. While both types of inquiries involve collecting data on the competition, they differ in their goals, methods, and, most importantly, their legality.[15]

Simply put, industrial espionage is illegal, and apprehension can result in severe consequences for the company that employs the spy as well as the individual(s) involved in the operation. Methods used in industrial espionage include hacking, false identities, bribery, and theft of physical objects and documents.

---

14. Kevin P. Riehle, "Russia's Intelligence Illegals Program: An Enduring Asset," in *Intelligence and National Security*, Volume 35, Issue 3, 2020, pp. 385–402.

15. John J. McGonagle and Carolyn M. Vella, *Proactive Intelligence: The Successful Executive's Guide to Intelligence* (London: Springer, 2012).

Business intelligence, in contrast, uses legal and ethical methods to collect information. This type of inquiry gathers information about competitors, customers, and the market, typically done through open-source research and other legal means.

Competitor intelligence entails systematically collecting, analyzing, and applying information about a company's competitors. The goal of competitor intelligence is to understand the strengths and weaknesses of a company's competitors and the opportunities and threats (i.e., a SWOT analysis) of its operations so that these insights can inform business decisions.

Employees can conduct competitor intelligence for their employer—these roles include market research analysts, strategic planners, and business intelligence professionals. Outside consultants or market research firms can also perform competitive intelligence.[16]

There are several methods used in competitor intelligence, including:

1. Primary research: This involves conducting direct research through surveys, focus groups, or in-depth interviews with customers, employees, and industry experts.

2. Secondary research: This involves gathering information from publicly available sources, such as annual reports, industry publications, and online databases.

---

16. Leonard M. Fuld, *The New Competitor Intelligence: The Complete Resource for Finding, Analyzing, and Using Information about Your Competitors, 2nd Edition* (New York: John Wiley & Sons, 1994).

3. Online monitoring: This involves using methods and techniques to monitor a competitor's online presence, including their website, social media, and other digital channels.

4. Competitive analysis: This involves analyzing a competitor's products, pricing, marketing strategies, and other business practices to gain a better understanding of their strengths and weaknesses.

5. Benchmarking: This involves comparing a company's performance against its competitors to identify areas for improvement and opportunities for growth.

So, business intelligence must not be confused with industrial espionage or associated with the latter's unethical means of gathering information. Companies that engage in business intelligence adhere to laws and ethical codes, thus avoiding activities that could harm their reputation or result in legal action.[17]

---

17. John J. McGonagle and Carolyn M. Vella, *Outsmarting the Competition: Practical Approaches to Finding and Outsmarting the Competition* (London: McGraw-Hill, 1993).

— CHAPTER TWO —

Typology of Industrial Espionage

A lthough the term *industrial espionage* may appear to refer to a monolithic phenomenon, this type of spying comprises different types of inquiry.[18]

Economic Espionage

Economic espionage is a type of spying that targets trade secrets and confidential business information. It occurs when a competitor, foreign government, or criminal organization seeks access to proprietary information such as research and development plans, customer lists, marketing strategies, or production processes.[19]

Sherman Kent described traditional economic espionage in his watershed study as a means of observing:

> ...new crops and the development of new methods of agriculture, changes in farm machinery, land use, fertilizers, reclamation projects, and so on. It must follow the discovery of new industrial processes, the emergence of new . industries, and the sinking of new mines. It must follow the development of new utilities and the extensions of those already established. It must follow changes in the techniques and implement of

---

18. Adam L. Penenberg and Marc Barry, *Spooked: Espionage in Corporate America* (Cambridge, MA: Perseus Publishing, 2000).

19. Mark Burton, "Government Spying for Commercial Gain," in *Studies in Intelligence*, Volume 37, Number 5, 1994, pp. 17–23.

distribution, new transport routes and changes in the inventory of the units of transportation.[20]

One of the most common forms of economic espionage is corporate espionage, which involves using covert means to gather information from a rival company. This can include hacking into computer systems, intercepting e-mails, bribing employees, or planting spies within the company. In some cases, the information gathered is used to create competing products, or it could be used to gain an advantage in negotiations or to undercut a rival's prices. In other cases, the information can be used to manipulate financial markets or to engage in insider trading, which will be discussed shortly.

Economic espionage is a concern for businesses and governments worldwide. "The U.S. Intellectual Property Commission estimated [in 2017] trade secret theft costs 1–3% of GDP, meaning that the cost to the $18 trillion U.S. economy was between $180 billion and $540 billion."[21]

In addition to financial losses, businesses can suffer damage to their reputation, making it difficult to attract new customers or retain existing ones. In some cases, companies can be forced into receivership, bankrupting them.

Scientific Espionage

Scientific espionage is another type of industrial spying focusing on proprietary scientific information. This type of espionage involves the theft of scientific research data,

---

20. Sherman Kent, *Strategic Intelligence for American World Policy* (Princeton, NJ: Princeton University Press, 1949), pp. 34–35.

21. Mark Button, "Editorial: Economic and Industrial Espionage," in *Security Journal*, Issue 33, 2020, pp. 1–5.

such as developing new technology and creating innovative goods or services or ground-breaking scientific findings.

Scientific espionage can take various forms, including hacking into computer systems, bribing employees, and infiltrating organizations through fake job applications.

Political Espionage

Political espionage is collecting sensitive information by government or political organizations to influence or undermine their policies, decisions, and actions, often using the same "...means, methods and practices."[22] Espionage of this nature is carried out by agencies like the Chinese Ministry of State Security, the North Korean Bureau 121,[23] and the Russian Federal Security Service (FSB). It can range from gathering economic and trade information, but primarily, the stealing of classified political information.

Political espionage can also be used to gain insight into a foreign country's stability and monitor the actions of foreign leaders considered threats.

But political espionage is not confined to nation-versus-nation affairs. Take, for example, the case involving the June 17, 1972, break-in of the U.S.

---

22. Sabastian Becker, "Economic Espionage in the Early Modern Period," in Andreas Bähr, Guido Braun, Marion Gindhart, Susanne Lachenicht , editors, *Spies, Espionage and Secret Diplomacy in the Early Modern Period* (Kohlammer Verlag, 2021), p. 45.

23. Bureau 121 is reported to be North Korea's cyberwarfare agency, which is part of the country's larger Reconnaissance General Bureau.

Democratic headquarters in the Watergate building in Washington, D.C.[24]

The break-in was carried out by five men caught in the act and later linked to President Richard Nixon's re-election campaign.[25] The incident sparked an enormous investigation that revealed a web of political spying, sabotage, and corruption within the Nixon administration, including using government agencies for political purposes, the attempt to cover up the break-in and related crimes, and interference with the investigation.

The so-called *Watergate Affair* led to the indictment and conviction of several high-ranking officials in the Nixon administration and ultimately to the resignation of President Nixon in August 1974.[26] Not surprisingly, the term *Watergate* has since become synonymous with political corruption.

Military–Industrial Espionage

Military-industrial espionage refers to the theft of military and defense industry data. Intelligence units that are part of a nation's military can carry out this type of espionage. Although, a nation's national security or foreign

---

24. See, E. Howard Hunt, *American Spy: My Secret History in the CIA, Watergate and Beyond* (New York: Wiley, 2007); G. Gordon Liddy, *Will: The Autobiography of G. Gordon Liddy* (New York: St Martins, 1980); and James Walter McCord Jr, *A Piece of Tape; The Watergate Story: Fact and Fiction* (Rockville, MD: Washington Media Services, 1974).

25. James McCord and E. Howard Hunt were ex-CIA, and G. Gordon Liddy was ex-FBI. Some might classify them as "rough," but certainly, they were low moral outliers regarding the ethical caliber of former government intelligence officers.

26. Becoming the first and only U.S. president to resign from office.

intelligence agency is likely to have some overlapping interest in military secrets. Classified military plans, weapons systems, manufacturing facilities, supply chain information, production output, and research and development into new military-aligned technologies are of prime importance.

Military-industrial espionage can also be used to gain insight into a foreign country's military leadership and its military's order of battle.[27]

Intellectual Property Theft

The stealing of trade secrets from businesses is generally classified as intellectual property theft. Although referred to as *theft*, it encompasses the unauthorized use or misappropriation of proprietary information, such as trade secrets, patents, trademarks, and copyrights. An example of industrial espionage involving intellectual property theft is the Huawei case.

In 2019, the Chinese company, Huawei, was alleged to have stolen trade secrets from T-Mobile, a U.S.-based telecommunications company. According to T-Mobile, Huawei employees stole a robot used for testing smartphones and took photographs of the device, which was proprietary technology. The case was settled out of court, with Huawei agreeing to pay a substantial settlement and implement measures to protect trade secrets.[28]

---

27. *Order of Battle* is an analysis of an opposition's number and types of military force as well as the quality and quantity of its troop training and their armaments. Knowing an opposition's Order of battle important to battlefield success.

28. Mike Dano, "T-Mobile wins $4.8M Ruling Against Huawei Over Alleged Theft of Smartphone-Testing Robot 'Tappy'," May

Nation-states can legislate what could be argued as intellectual property theft.[29] Take the example of Communist China, where it is a requirement for some businesses that want to conduct business there; they first must transfer their trade secrets (e.g., computer code). The World Economic Forum stated, "China has joined Russia in tightening the requirements placed on foreign companies to store information within national borders."[30]

". . . there is just no country that presents a broader threat to our ideas, our innovation, and our economic security than China." FBI Director Christopher Wray, January 31, 2023[31]

According to the US-China Economic and Security Review Commission in a report to Congress, China "...depends on industrial espionage, forced technology transfers, and piracy and counterfeiting of foreign technology as part of a system of innovation mercantilism."[32] By obtaining intellectual property

---

22, 2017. Available at: https://www.fiercewireless.com/wireless/t-mobile-wins-4-8m-ruling-against-huawei-over-alleged-theft-smartphone-testing-robot-tappy. Accessed: February 21, 2023.

29. Some critics might argue that such a requirement is legalized extortion.

30. World Economic Forum, "Weaponized AI, Digital Espionage and Other Technology Risks for 2017." Available at: https://www.weforum.org/agenda/2017/01/technology-risks-amplified-by-global-tensions/. Accessed February 17, 2023.

31. Stated in a speech at the Ronald Reagan Presidential Library and Museum, Simi Valley, California.

32. US–China Economic and Security Review Commission, *2012 Report to Congress*, 112th Congress, 2nd session (Washington, DC: Government Printing Office, 2012), p. 421.

illegally, China avoids the high research and development costs and the long lead time that some discoveries take.

At first blush, requiring a business to use the host nation's resources (e.g., cloud computing and storage facilities), but when one realizes that the Chinese Communist Party can require access to the data held in those facilities, it means that those data are no longer "secret."[33] A company's lawyer should be consulted on how to engage in such contracts, *if at all*.

Sabotage

Even though sabotage sounds more like a war-like event,[34] sabotage against competitors forms part of industrial espionage; it includes any actions taken to damage competitors, thus gaining an advantage in the marketplace. And, as we saw with other types of espionage, it aims to achieve a market advantage that locates sabotage within this gambit.

Some methods used in business sabotage include spreading false information or rumours about the competitor, cyber-attacks, and manipulating the market by controlling supply and demand. Other techniques include interfering with a competitor's distribution networks, disrupting their production processes, and using lobbying and, ironically, legal tactics to restrict the competitor's operations. Bearing these approaches in mind, a study also found that industrial sabotage may "…have more

---

33. Clive Hamilton and Mareike Ohlberg, *Hidden Hand: Exposing How the Chinese Communist Party is Reshaping the World* (Richmond, Victoria: Hardie Grant Books, 2020).

34. See, for example, M.R.D. Foot, *SOE in France: An Account of the Work of the British Special Operations Executive in France 1940–1944, Revised Edition* (London: Frank Cass, 2004).

symbolic and moral power than it has instrumental or significant effect."[35]

An example of physical sabotage is the Nord Stream pipeline network that occurred on September 26, 2022. Nord Stream 1 (NS1) and Nord Stream 2 (NS2)—two pipelines comprising two pipes each—were built to carry natural gas from Russia to Germany via the Baltic Sea. The major owner is the Russian state-owned gas company, Gazprom.

Saboteurs cut the gas flow by detonating explosive devices that severed NS1 and NS2. Although the explosions occurred in international waters, they were within the economic zones of Sweden and Denmark; both countries launched investigations. Russia denied involvement, but some observers pointed out the parallel between this event and Russia's destruction of a pipeline to Georgia in 2006 when Georgia expressed interest in joining NATO.[36]

The pipelines were on operating at the time because of disputes between Russia and the European Union due to Russia's criminal invasion of Ukraine. But long before the sabotage, its building was controversial, with the U.S. criticizing it because Russia could use it to hold Europe "hostage."

At the time of writing, there was no clear indication of who might be responsible—Russia, one of its affiliates, a

---

35. Steve Linstead, "Breaking the 'Purity Rule': Industrial Sabotage and the Symbolic Process," in *Personnel Review*, Volume 14, Number 3, 1985, pp. 12–19.

36. Interestingly, Sweden and Finland applied to join NATO in July 2022, two months prior to the severing of the Nord Stream pipelines. Denmark was one of the founding members of NATO.

NATO member, or one of its associates. Whoever was responsible, carried out the mission in total secrecy—as an industrial sabotage mission dictates—and the result was as successful in eliminating the pipeline as a means of supplying Europe with Russian gas.

Another example of sabotage is a campaign of "dirty tricks."[37] Take the case of the 1972 U.S. presidential campaign. A political operative for the re-election of Richard Nixon was lawyer Donald Segretti. He perpetrated several covert operations[38] aimed at disrupting the Democratic Party's candidates.

Segrettis' operations involved various illegal and unethical activities—known as "opposition research"—including forging embarrassing letters, distributing fake campaign literature, and planting false stories in the media.[39]

The goal of the dirty tricks campaign was to undermine the credibility of Democratic Party candidates, thus helping to secure a victory for incumbent President Nixon. However, history shows that the dirty tricks program unravelled, leaving a connecting trail to the Watergate scandal.

Insider Trading

Insider trading is where an individual with privileged access to confidential information uses that information

---

37. Richard Ells and Peter Nehemkis, *Corporate Intelligence and Espionage* (New York: Macmillan Publishing Company, 1984), pp. 126–127.

38. These operation were dubbed "ratfucking" by Segretti.

39. Tony Ulasewicz with Stuart A. McKeever, *The President's Private Eye: The Journey of Detective Tony U. from NYPD to the Nixon White House* (Westport, CT: MACSAM Publishing, 1990).

for personal financial gain. This is often seen in media reports where a person has bought and sold stocks based on non-public information. This is highly illegal and can result in serious consequences.[40]

A notable insider trading case is that of Mr Raj Rajaratnam, the founder of the Galleon Group hedge fund. In 2011, Mr Rajaratnam was convicted of securities fraud and conspiracy for insider trading based on tips from corporate insiders, analysts, and others. He was sentenced to eleven years in prison and ordered to pay a $US10 million fine.[41]

The other case worth noting is that of Ms Martha Stewart, the founder of Martha Stewart Living Omnimedia. In 2004, she was convicted of conspiracy, obstruction of an agency proceeding, and making false statements to federal investigators about ImClone Systems stock insider trading. She served five months in federal prison and was ordered to pay a $US30,000 fine.[42]

An international case involved a Russian national named Yevgeniy Nikulin, who was convicted of hacking into the computer systems of several U.S. companies, including *LinkedIn*, *Dropbox*, and Formspring. Nikulin

40. Nick Catrantzos, *Managing the Insider Threat: No Dark Corners* (Boca Raton, FL: CRC Press, 2012).

41. Brian Duignan, "Raj Rajaratnam." Available at: https://www. britannica.com/biography/Raj-Rajaratnam. Accessed February 21, 2023.

42. "Martha Stewart Living Omnimedia, Inc." Available at: https://www.encyclopedia.com/social-sciences-and-law/ economics-business-and-labor/businesses-and-occupations/ martha-stewart-living-omnimedia-inc. Accessed February 21, 2023.

was arrested in Prague in 2016 and extradited to the U.S. in 2018 to stand trial on hacking and identity theft charges.

During the trial, evidence showed Nikulin had used spear-phishing attacks to gain access to the e-mail accounts of *LinkedIn* and Formspring employees. He then used this access to steal user credentials, which he later sold on underground marketplaces. The stolen information was also used to access user accounts on *LinkedIn* and *Dropbox*, where Nikulin stole sensitive information.

The case gained added significance due to allegations that Nikulin had ties to the Russian government. U.S. officials had long accused the Russian government of engaging in cyberattacks against American targets, and some saw Nikulin's arrest as an opportunity to shine a light on these alleged activities.

While the evidence presented at trial did not definitively prove that Nikulin was working on behalf of the Russian government, it did suggest that he had connections to Russian intelligence agencies. The prosecution argued that Nikulin boasted about his links to Russian intelligence and had even offered to work for them.

In the end, Nikulin was found guilty of multiple hacking and identity theft counts and sentenced to eighty-eight months in prison.[43]

---

43. Tom Winter and Ken Dilanian, "Russian National with Ties to Putin Convicted of Hacking U.S. Companies in $90 Million Insider Trading Scheme." Available at: https://www.nbcnews.com /politics/russian-national-ties-putin-convicted-hacking-us-companies-90-million-rcna70652. Accessed February 21, 2023.

Some of the more common methods used to commit insider trading[44] include:

1. Tipping: Sharing privileged information with others who use it to make trades.

2. Front-running: Using privileged information to make trades ahead of a planned transaction by the company, such as a merger or acquisition.

3. Trading on material non-public information: Using privileged information about a company's financial results, product launches, or other significant events that have not been made public to make trades.

4. Misuse of confidential information: Using confidential information obtained through employment to make trades.

5. Parallel trading: Making trades similarly to others with access to privileged information.

---

44. Insider trading involves the illicit use of non-public, material information to gain an unfair advantage in trading securities. Legal and regulatory frameworks aim to maintain market integrity by ensuring all market participants have equal access to material information.

— CHAPTER THREE —

Myths and Realities

According to Bruce Wimmer, a variety of misconceptions undermine the safeguarding of confidential business information.[45] He outlines four that he describes as the *Silo Syndrome*, where information is compartmentalized and isolated within different segments of an organization, hindering effective communication and collaboration. The *James Bond Syndrome* also reflects an overemphasis on espionage and external threats, potentially neglecting internal vulnerabilities. The *Ostrich Syndrome* denotes a scenario where individuals or organizations ignore the potential risks and threats to data security, hoping they will not be affected.

Furthermore, there is also a tendency for an excessively focused approach to cyber-security, which, while important, may lead to the neglect of other essential aspects of data protection, such as physical security and employee training. The widespread nature of these four issues suggests a need for a thorough examination of the common challenges that hinder effective data protection strategies.

Wimmer reported that experts specializing in counterespionage identified the Silo Syndrome as a paramount concern for entities grappling with corporate

---

45. Bruce Wimmer, *Business Espionage: Risks, Threats, and Countermeasures* (Amsterdam: Elsevier, 2015), pp. xv–xviii.

spying.[46] The syndrome is characterized by the tendency within organizations to delegate espionage-related issues to specific, isolated departments such as security, IT, legal, or human resources, thus pigeonholing it as a problem exclusive to those areas. This compartmentalization leads to a narrow perspective on espionage, often summed up in assertions like "This is an issue for the security team" or "This falls under the purview of IT."

The expert panel at the conference emphasized a crucial shift in perspective: recognizing business espionage as a comprehensive organizational challenge rather than a segmented departmental issue. They advocated a unified approach transcending conventional departmental divisions, calling for leadership to navigate these internal barriers. Such leadership would bridge the gaps between organizational functions and foster a collective strategy against corporate espionage. This approach necessitates a figure or figures within the organization who wield sufficient authority and capability to orchestrate a cohesive response across the entire corporate structure, thereby ensuring a robust and integrated defense against espionage threats.

A prevalent but mistaken belief exists that industrial, corporate, or business espionage is exclusively a high-tech crime carried out by individuals reminiscent of James Bond. These imagined scenarios often involve dramatic infiltrations, such as descending into office buildings or manufacturing sites through air conditioning ducts on

---

46. Bruce Wimmer, *Business Espionage: Risks, Threats, and Countermeasures*, p. xv.

special wire cables. Alternativ[47]ely, crime is considered the domain of highly skilled computer hackers, embodying the genius yet socially awkward nerd stereotype. However, both representations are far from accurate. As subsequent discussions reveal, most corporate espionage is executed through surprisingly straightforward and preventable techniques.

With this misunderstanding and inadequate defenses against business espionage, Wimmer suggests that even someone as inept as Maxwell Smart, a character from the 1960s American television series "Get Smart" known for his bumbling incompetence, could successfully exfiltrate valuable corporate secrets.[48] The comparison underscores a critical point: the real-world vulnerability of many businesses to espionage lies not in failing to thwart the sophisticated tactics of a cinematic super-spy but in overlooking the risk posed by much more straightforward methods of information gathering. This gap in understanding and protection exposes businesses to significant risks, emphasizing the need for a more comprehensive approach to safeguarding against espionage.

Many business leaders, including those responsible for information management and security, harbor the unfounded optimism that their organizations are unlikely to be targeted by espionage. This belief is often rooted in wishful thinking rather than a rigorous analysis of facts. It mirrors the metaphorical behavior of an ostrich burying its

---

47. Bruce Wimmer, *Business Espionage: Risks, Threats, and Countermeasures*, p. xv.

48. "Get Smart" was a satirical comedy that parodied the spy genre, and Maxwell Smart, or Agent 86, was its haplessly ineffectual protagonist.

head in the sand at the first sign of danger under the misguided assumption that invisibility equates to safety. There is a prevailing assumption among such executives that unless their company operates within the defense sector, possesses highly technical knowledge, or is of a considerable size, it is somehow immune to the threats of corporate espionage. A common yet erroneous belief is articulated: "Our company possesses no assets worth taking, or our technology advances at such a pace that it would be outdated by the time someone could steal it." This mindset inadvertently presents spies with prime opportunities for exploitation.

In reality, small businesses frequently become targets more often than their larger counterparts do, not only due to their sheer number and the increased likelihood of competition but also because they generally implement less stringent security measures. The notion that any company could be exempt from the interest of corporate spies is a dangerous underestimation. For a small enterprise, a financial loss of, say, $75,000 can have a far more profound impact than a loss of hundreds of millions of dollars on a large corporation. If an organization possesses nothing of value worth stealing (and thus protecting), it might well reflect a lack of competitiveness within its industry.

Wimmer stated that this dismissive attitude often comes to light during security risk assessments.[49] However, upon discussing the nature of the business, its competitive advantages, and future objectives, it becomes evident that these organizations possess valuable trade secrets, even if they have not formally recognized them.

---

49. Bruce Wimmer, *Business Espionage: Risks, Threats, and Countermeasures*, p. xv.

This oversight highlights potential gaps in legal protection for these secrets and indicates a lack of adequate measures to safeguard sensitive information.

Another dimension of this so-called ostrich syndrome manifests in the eagerness to expedite changes and implementations, particularly in relocating manufacturing operations abroad. Companies often prioritize speed over security, adopting an "ignore any wrongdoing and turn a deaf ear to any misconduct" stance that willfully disregards potential risks. This approach can lead to catastrophic decisions without considering the threats of industrial espionage and the loss of intellectual property, underscoring the critical need for comprehensive risk assessment and protective strategies in the decision-making process.[50]

The information technology sector has made significant strides in addressing the issue of intellectual property loss, leading some specialists to prioritize cyber-security as the primary defense against business espionage mistakenly. However, it is crucial to recognize that valuable information is not confined to digital formats. Regardless of its storage medium, a piece of information holds equal value; whether digital data on a computer or notes scribbled on a piece of paper, both forms are equally enticing to industrial spies. These individuals often collect fragments of information, piecing them together to form actionable intelligence. Thus, it is imperative to safeguard all types of sensitive business information, irrespective of the format in which it is stored.

---

50. Bruce Wimmer, *Business Espionage: Risks, Threats, and Countermeasures*, p. xv.

Sensitive information encompasses various data types, including digital files, paper documents, photographs, observations, and verbal communications. This information can range from formal documents to drafts, working papers, internal communications, and casual conversations in both formal and informal settings.[51] While cyber-security is an indispensable component of any modern counterespionage strategy, focusing exclusively on digital data protection leaves organizations vulnerable to many traditional espionage tactics.

The primary objective of industrial espionage is acquiring "information"—any knowledge that could potentially harm your organization or benefit a competitor. According to law enforcement agencies, industrial espionage inflicts financial losses of billions of dollars annually. Technical vulnerabilities, particularly a cyber vector, account for, perhaps, as little as a quarter of all instances where sensitive business information is compromised.[52] However, the threat of cyber espionage is on the rise.

Despite the critical importance of cyber-security, it is a grave mistake for businesses to channel all their counterintelligence efforts into the digital domain alone. Such a narrow focus neglects up to three-quarters of potential threat vectors. Business spies often prefer the path of least resistance, targeting the most accessible and, consequently, most minor protected sources of information. These sources include discarded documents,

---

51. Henry W. Prunckun, *Information Security: A Practical Handbook on Business Counterintelligence* (South Australia: Bibliologica Press, 2020).
52. This is an estimate. The true magnitude is, or course, not known.

unsecured telephone conversations, or overly talkative employees. These methods are less risky for the spy and often go unnoticed, making them an attractive first option before resorting to more sophisticated or high-tech approaches.

Moreover, exploiting IT security can sometimes involve simple tactics like using a pretext to obtain passwords or leveraging physical security weaknesses to gain direct access to servers or communication lines. In some cases, it involves placing a spy within the company or exploiting an insider with legitimate access to IT systems. The diversity of these methods underscores the necessity for a comprehensive approach to counterintelligence encompassing a wide array of protective measures beyond just cyber-security.

Case Study

According to the FBI, in January 2012, Wen Chyu Liu, a retired research scientist from Dow Chemical Company, received a sentence of 60 months in prison accompanied by two years of supervised release, a financial penalty of a $25,000 fine, and an order to forfeit $600,000. This sentencing followed Liu's conviction in February 2011 for illegally acquiring and selling proprietary information related to Dow's elastomers Tyrin CPE process and product technology, intending to benefit entities in China.[53]

Engaging in a collaborative scheme with at least four individuals, both current and former employees, Liu undertook extensive travels across China to disseminate

---

53. Center for Development of Security Excellence, *Insider Threat: Wen Chyu Liu* (Washington, DC: Defense Counterintelligence and Security Agency, n.d.).

the illicitly obtained information. He engaged in financial transactions with these employees, compensating them for proprietary materials and information. Furthermore, Liu was found to have provided a cash bribe of $50,000 to an employee to acquire a process manual, among other confidential documents.[54]

Behavioral Indicators

According to the FBI, certain behaviors might signal an employee engaging in espionage or systematically misappropriating resources from their organization.[55] These indicators include the unauthorized removal of confidential or proprietary materials from the workplace, whether through physical documents, electronic storage devices like thumb drives and computer disks, or email. Employees might also display an undue interest or attempt to access sensitive information unrelated to their job responsibilities or show curiosity in topics that could be of value to foreign entities or competitors.

Copying materials without necessity, particularly those classified or proprietary, or accessing the organization's computer network during non-working hours, such as vacations, sick leaves, or unusual times, could also be warning signs. Another red flag is disregarding organizational policies on using computer resources, such as installing unauthorized software or hardware, visiting

---

54. Federal Bureau of Investigation, "Former Dow Research Scientist Sentenced to 60 Months in Prison for Stealing Trade Secrets and Perjury," *Press Release* (Washington, DC: Department of Justice, January 13, 2012).

55. Federal Bureau of Investigation, *The Insider Threat* (Washington, DC: Department of Justice, n.d.), p. 2.

restricted websites, conducting prohibited searches, or downloading confidential data.

Additional concerning behaviors include working unusual hours without permission, a keen interest in working overtime or during weekends on schedules that might facilitate covert activities, unreported interactions with foreign nationals potentially linked to foreign governments or intelligence agencies, or undisclosed travel abroad. Short, unexplained trips to foreign countries, an inexplicable increase in wealth beyond what their income level would typically support, or engaging in suspicious interactions with competitors, business partners, or others not authorized to receive company information are notable indicators.

An employee may also exhibit signs of being under significant personal or professional stress, show an unusual curiosity about the personal affairs of colleagues, or exhibit paranoia about being under surveillance, possibly going as far as setting up traps to check for searches of their personal spaces or looking for hidden surveillance equipment.

However, the FBI has pointed out that it is important to note that many individuals might exhibit some of these behaviors to various extents without ever engaging in illegal activities.[56] The presence of one or more of these behaviors *does not* necessarily confirm wrongdoing. However, it may warrant closer attention or investigation to ensure the security and integrity of the organization's assets and confidential information.

---

56. Federal Bureau of Investigation, *The Insider Threat*, p. 2.

— CHAPTER FOUR —

Methods of Industrial Espionage

A s with all types of espionage, information collection is not a haphazard process. The information needed to answer decision-makers' questions, by the nature of the questions, is specific. These specifications are referred to as *intelligence requirements*. We can think of intelligence requirements as a shopping list of data items analysts need to draw conclusions about.[57] In turn, these conclusions help decision-makers make their plans.

It is important to understand that intelligence analysts "draw conclusions" based on fact and logic but do not "predict." The best way to view an intelligence report is that its conclusions *reduce uncertainty*. Uncertainty cannot be eliminated.

That said, many methods exist for obtaining confidential information to supply the analytic process. Still, these varying methods can be categorized into a few groups, which we will discuss here—surveillance, infiltration, exploitation of human sources, and cyber penetration. But first, we will examine the types of businesses that spies "target."

Arguably, the type of business that is most exposed to industrial espionage is that that is operating in the field of

---

57. Brian Manning and Kristan J. Wheaton, "Making 'Easy Questions' Easy: The Difficulty of Intelligence Requirements," in *International Journal of Intelligence and Counterintelligence*, Volume 26, Number 3, 2013, pp. 597–611.

innovation—regardless of whether it is technological, medical, or scientific discoveries.

Other attractive targets are companies with government, defence or security contracts. Why? Because these businesses are at the lead for changes in goods and services that will not only make (potentially) a lot of money, but the research and development required to bring about these discoveries carry substantial costs. It is self-evident that if one can steal these secrets, it will save time and money trying to keep up with competitors. This logic applies to government policy (e.g., international relations) and the defense sector regarding foreign adversaries.

According to United States federal law, *trade secrets* are described as "...all forms and types of financial, business, scientific, technical, economic, or engineering information, including patterns, plans, compilations, program devices, formulas, designs, prototypes, methods, techniques, processes, procedures, programs, or codes, whether tangible or intangible, and whether or how stored, compiled, or memorialized physically, electronically, graphically, photographically, or in writing."[58]

How does the industrial spy select a method of acquisition? It is chosen by where the data resides. The most direct route to that information is the most efficient but may also be fought with risk. So, an indirect approach

---

58. U.S. Code, Title 18, Part I, Chapter 90 § 1839

may be what's needed. That is, data access via a third party or a stepping-stone technique.

For example, a small, unimportant business may service a larger company, which is the target. This smaller business may be penetrated first to reach the prime target.

Surveillance

We will discuss two approaches to surveillance—electronic and physical. Companies are increasingly using electronic surveillance to monitor the performance of their employees. While this surveillance aims to ensure productivity for the most part,[59] it also has a secondary role in protecting companies from espionage committed by the same employees.

Establishing ways to track employees' movements in the workplace or the information they accessed from the company's databases is a defensive mechanism that can discourage efforts to collect confidential information. This approach is a *defensive* form of electronic surveillance; it includes CCTV cameras, computer logs, and ID access cards.[60] It is termed *security intelligence*, and its function comes under the umbrella of business *counterintelligence*.

The other approach to electronic surveillance is *offensive*, and it, too, is used by industrial spies. Offensive electronic surveillance is used to collect information from, say, social media to identify company personnel and their

---

59. Jonathan P. West and James S. Bowman. "Electronic Surveillance at Work: An Ethical Analysis," in *Administration & Society*, Volume 48, Number, 2016, pp. 628–651.

60. Henry W. Prunckun, *Information Security: A Practical Handbook on Business Counterintelligence*, 2019.

habits and traits to gauge behaviours exhibited in their public socio-political views.

Physical surveillance involves following a person to determine what they are doing, who they are meeting with, and establishing their routines. The benefit of physical surveillance over cyber surveillance is the ability to ascertain what is real and portrayed online. There is a well-identified schism between how people show themselves on social media and in reality.[61] Physical surveillance can provide a more realistic understanding of the person under consideration.

Nevertheless, physical surveillance is expensive and time-consuming. It is safe to say that unless the information sought through physical surveillance outweighs the cost, a spy will likely seek other avenues.

Although we have discussed physical surveillance regarding a person following another person, "following" can be done via distance techniques, such as using a drone, telescope, binoculars, telephoto camera, or tracking device on the target's vehicle. Although, in the latter's case, a tracking device could be considered a type of electronic surveillance.[62]

---

61. A. Lieberman and J. Schroeder. "Two Social Lives: How Differences between Online and Offline Interaction Influence Social Outcomes, in *Current Opinion in Psychology*, Volume 31, February 2020, pp. 16–21.

62. For an in-depth discussion of physical surveillance, we recommend Henry W. Prunckun, *Espionage Black Book Three: Surveillance Explained* (South Australia: Bibliologica Press, 2021).

Infiltration and Pretexts

Infiltration is related to police undercover work. However, unlike police operatives who operate within the law, industrial spies do not. Physical infiltration involves the spy entering the target company. Depending on how this is done, it might amount to criminal or civil trespass.

It may be the case that a spy uses the guise of a genuine employee or a contractor. Or, it might be the case of a person seeking employment for the sole purpose of stealing secrets or an existing employee who has been recruited as an agent.

A notable example is the case of Xiaoqing Zheng an engineer working for General Electric in the United States. Zheng was sentenced to twenty-four months in prison for conspiring to steal General Electric's trade secrets for the People's Republic of China (PRC). He was employed with GE for approximately ten years and conspired to steal ground-based and aviation-based turbine technologies.[63]

Physical infiltration is a low-tech approach to collecting information, yet it is a high-risk challenge. Still, with the ability to access classified information via the Internet, physical infiltration is sometimes overlooked.

The use of *security intelligence* and its associated methods help reduce physical infiltration to keep a from being penetrated. Businesses need to understand the sensitivity of the information they possess so that data

---

63. U.S. Department of Justice, "Former GE Power Engineer Sentenced for Conspiracy to Commit Economic Espionage," *Press Release*, dated January 3, 2023. Available at: https://www.justice.gov/opa/pr/former-ge-power-engineer-sentenced-conspiracy-commit-economic-espionage. Accessed March 15, 2023.

managers can assess the level of risk to each information item.[64]

In this regard, some security consultancies provide "red teaming" services. This approach entails operatives covertly testing the business's physical security and how the company has compartmentalized various classes of information.

Red teaming involves, say, a person playing the part of a spy providing a false reason for being present on the business's premises, such as an appointment with a staffer working in the building or that they have been contracted to undertake work (e.g., as a cleaner, or a tradesperson, like an electrician). Infiltration of this type uses what is termed a *pretext*.[65]

A pretext is a story—a justification for being in a particular place or making inquiries. It is used to ease suspicion. Good pretexts will sound realistic but not overly complicated. A pretext may also be in the guise of a customer inquiry. Questions about a product, availability, cost, or origin can provide initial data that can be used for planning the penetration-in-chief, which comes later.

Human Source Recruitment

Developing a human source for intelligence is an art practiced by law enforcement investigators, intelligence operatives, and undercover military personnel. The principles used in these sectors have been adapted to industrial espionage. "Anyone who knows the techniques

64. M. Podszywalow, "Preventing Corporate Espionage," in *Risk Management*, Volume 59, Number 2, 2012, pp. 24–26.

65. Greg Hauser, *The Pretext Manual*, 1994.

that case officers working for the CIA, KGB, MI6, and Mossad used to recruit spies can use those same techniques to operate an intelligence-collection operation against any business government agency, political organization, or individual."[66]

The time-proven approach to developing a human intelligence source was called *MICE*—a mnemonic that identifies a person's motives, which are then used as possible avenues for exploitation. The letters stand for money, ideology, compromise, and excitement. These human motivation categories are viewed as recruiting vectors (those recruited are known as *agents*[67]).

By understanding and leveraging these motivations, industrial spies can target individuals susceptible to recruitment and use these motivational weaknesses to convince them to cooperate. Let us look at each of these elements.

Money refers to the use of financial incentives to offer the recruit. The promise of a large sum of money or ongoing financial support can be a powerful motivator for individuals undergoing financial stress.

Ideology refers to the use of, say, religious or political motivations in the recruitment process. For example,

---

66. Jefferson Mack, *Running a Ring of Spies: Spycraft and Black Operations in the Real World of Espionage* (Boulder CO: Paladin Press, 1996), p. .33.

67. An *agent* is a person who acts on behalf of another, in this context, a *case officer*. Not to be confused with an FBI officer, who is referred to in the stylized-address as an *agent*. Bob Burton, *Top Secret: A Clandestine Operator's Glossary of Terms* (Boulder, CO: Paladin Press, 1986), pp. 6–7.

someone may be motivated to spy because they believe in the ideology or cause that the recruiter represents.

Coercion refers to the use of threats or blackmail. Someone may be blackmailed with compromising information or threatened with harm to themselves or their loved ones if they do not cooperate.

Ego refers to the use of flattery or appeals to someone's self-esteem to recruit them as an agent. If the person can be convinced that they are uniquely qualified or important to the mission's success.

It is important to note that the use of the MICE approach has stood the test of time—dating back to the days of the U.S. Office of Strategic Services[68]— nonetheless, a different approach has since emerged— RASCLS. Like MICE, this mnemonic stands for revenge, ideology, sex, coercion, loyalty, and sympathy). Beckett states, "Although MICE provides superficial explanations for spying, it fails to capture the complexities of human motivation."[69]

Burkett argues that using Cialdini's[70] six principles of persuasion framework (RASCLS), industrial spies are able "…to see opportunities to find and recruit agents from

68. The Office of Strategic Services (OSS) was founded on June 13, 1942.

69. Randy Burkett, "An Alternative Framework for Agent Recruitment: From MICE to RASCLS," in *Studies in Intelligence*, Volume 57, Number 1, March 2013, p. 11.

70. Robert Cialdini, *Influence: The Psychology of Persuasion* (Quill/William Morrow, 1984).

a population beyond those defined by the vulnerabilities exploitable in the MICE framework."[71]

Here is an explanation of each of these six principles:

1. Reciprocation: People feel obliged to repay others for what they have received. For example, if a recruiter gives a person a gift, they may feel obligated to return the favor in the future.

2. Authority: People are more likely to comply with requests from those who are perceived to be experts or authorities in a particular field. If the recruiter presents an air of power, the person may be more likely to take participate in the secret operation.

"As Victor Ostrovsky, the ex-Mossad case officer, explained, 'The idea of recruitment is like rolling a rock down a hill ... you take someone and get him gradually to do something illegal or immoral. You push him down the hill. We didn't blackmail people. We didn't have to. We manipulated them.'"[72]

---

3. Scarcity: People value things that are perceived to be rare or in short supply. If the recruiter presents the person with a "fleeting opportunity," they may feel the urgency to participate.

4. Commitment and consistency: People have a desire to be consistent with their past actions and statements. If recruiter can get the person to commit to a position,

---

71. Randy Burkett, "An Alternative Framework for Agent Recruitment: From MICE to RASCLS," 2013, p. 13.

72. Cited in Jefferson Mack, *Running a Ring of Spies*, 1996, p. 47.

that person is more likely to follow through with supporting that position to maintain consistency.

5. Liking: People are likelier to comply with requests from those they like or find attractive. If the recruiter is friendly and engaging, shows shared interests, and demonstrates they can be a "sounding board" or confident, they will be more successful in persuading a person to obtain information for them.

6. Social proof: People are more likely to follow the actions of others in similar situations. If the recruiter can convince the person that "others have provided similar information," the person may be inclined to do the same, the way a person might join a line at a restaurant, thinking it is popular, so it must be good.

Cyber Penetration

Cyber espionage has become commonplace. Broadly, "…cyber espionage, or cyber spying, is a type of cyber-attack in which an unauthorized user attempts to access sensitive or classified data, intellectual property (IP) for economic gain, competitive advantage, political reasons or military advantage."[73] Those conducting cyber espionage are "the usual suspects"—business competitors, hacktivists, and state-sponsored foreign attackers.[74]

---

73. Shkurte Luma-Osmani, Abdulla Civull, Gjulie Arifl, and Eip Rufati, "Cyber Espionage Consequences as a Growing Threat," in *Journal of Natural Sciences and Mathematics of University of Tetova*, Volume 7, 2002, pp. 13–14.

74. PriceWaterhouseCoopers, *Study on the Scale and Impact of Industrial Espionage and Theft of Trade Secrets through Cyber*, European Union, 2019. Available at: https://www.pwc.com/it/it/publications/docs/study-on-the-scale-and-Impact.pdf. Accessed March 27, 2023.

We can see how state-sponsored industrial spies used the Internet to penetrate businesses in Operation Night Dragon. This was a hacking operation that was discovered in 2011. It targeted several major energy companies around the world. The campaign is considered to have been sponsored by a Chinese hacking group, "Comment Crew," which had previously been linked to other high-profile cyber-attacks.[75]

The goal of Operation Night Dragon was to steal sensitive information from the energy companies, including information related to oil and gas field exploration, procurement, and financial data. The attackers used various methods to access the targeted networks, including spear-phishing e-mails, malware-laden documents, and applications that would allow secret remote access.

Once inside the network, the industrial spies used several techniques to gather information and maintain access to the systems. The discovered applications included keylogging programs, password theft applications, and programs that allowed the spies remote administration. The stolen data was then exfiltrated to servers under the attackers' control.

The Operation Night Dragon operation is notable because of its persistence and the sophistication of the techniques used. It was also one of the first significant cyber-attacks to target the energy sector, a critical infrastructure industry. The campaign raised concerns about the vulnerability of critical infrastructure to cyber-

---

75. "Comment Crew" is nickname for the People's Liberation Army Unit 61398 stationed in Pudong, Shanghai.

attacks and led to increased investment in cyber-security for the energy industry.

The methods used in this approach are, of course, reliant on the fact that the information is on a computer system. If the information is not online, other infiltration forms are required. For this reason, cyber penetration has advantages because it does not need an operative to be physically at the target location. This reduces the risk of identification and increases the uncertainty of the information taken.

Suppose a business cannot determine the data taken because the penetration and the spy are unknown. In that case, the company cannot concentrate on "damage control."[76] Therefore, it must be assumed that all data was taken, which drives up the resources required and increases the time and expense needed to mitigate the harm that will result.

Cyber espionage need not be complicated or high-tech. Granted, there are cases where complex software applications and risky tradecraft have been used to install these programs. However, it could also be as simple as obtaining someone's password or user account details.

---

76. *Damage control* refers to the actions taken to limit the negative impact of mistakes, crises, or scandals on their reputation, financial stability, or operational capacity. This process involves assessing the extent of the damage, communicating with stakeholders, and implementing measures to mitigate the consequences. The aim is to restore trust, maintain credibility, and prevent further harm by addressing the cause(s) of the issue and taking steps to ensure similar incidents do not recur. Effective damage control requires timely intervention, fast planning, and, often, a commitment to change to regain the confidence of affected parties.

A common tactic used in cyber espionage is using different versions of *malware*. Malware is a collective name of a group of malicious software programs. They go by names that reflect their operations: worms, rootkits, spyware, and Trojans.[77]

Worms. A worm malware program is a type of self-replicating malicious software that spreads across computer networks and systems by exploiting security vulnerabilities. Unlike viruses, worms do not require a host program or file to infect other computers. Instead, they can spread independently by exploiting network or system vulnerabilities.

Once a worm infects a computer or network, it can replicate itself and spread to other computers, often consuming network resources and causing system slowdowns or crashes. Worms can also perform malicious actions, such as stealing data, creating backdoors for remote access, or launching DDoS (Distributed Denial of Service) attacks on targeted websites or servers.

Because worms can spread quickly and autonomously, they can be dangerous and difficult to contain. It is important to keep software and operating systems up to date with the latest security patches and to use reputable antivirus software to prevent worm infections.

Rootkits. A rootkit malware program is malicious software that allows unauthorized users to gain access and control over a computer system or network while remaining undetected by traditional security measures.

---

77. Gaute Wangen, "The Role of Malware in Reported Cyber Espionage: A Review of the Impact and Mechanism," in *Information*, Volume 6, Number 2, 2015, pp. 183–211.

Rootkits are designed to hide their presence from the user and the operating system, making them difficult to detect and remove.[78]

Rootkits can be used maliciously, such as stealing sensitive data, modifying system settings, or installing additional malware. They typically gain access to a system by exploiting software vulnerabilities or tricking users into downloading and installing them. Once installed, a rootkit can modify system files and settings to hide its presence and establish a backdoor for remote access by the attacker.

Rootkits can be challenging to detect and remove because they operate low within the operating system and can evade traditional antivirus and anti-malware programs. Detecting and eliminating rootkits often requires specialized tools and techniques. To protect against rootkits, it is important to keep software and operating systems up to date with the latest security patches and to use reputable antivirus and anti-malware software.

Spyware. Spyware is malicious software designed to secretly monitor and gather information from a user's computer or mobile device. The information collected can include keystrokes, browsing history, passwords, and other sensitive data, which can be used for nefarious purposes such as identity theft, financial fraud, or corporate espionage.[79]

---

78. Christopher C. Elisan, Malware, *Rootkits and Botnets: A Beginner's Guide* (New York: McGraw-Hill Education, 2012).

79. Laurent Richard and Sandrine Rigaud, *Pegasus: The Story of the World's Most Dangerous Spyware* (New York: Macmillan, 2023).

Spyware typically enters a system through malicious e-mail attachments, software downloads, web browsers, and other software vulnerabilities. Once installed, it can run in the background without the user's knowledge, sending the gathered information to remote servers controlled by the attacker.

In addition to monitoring user activity, spyware can modify system settings and install additional malware. Spyware can be difficult to detect and remove because it is often designed to evade traditional antivirus and anti-malware programs.

To protect against spyware, it is important to use reputable antivirus and anti-malware software and to keep software and operating systems up to date with the latest security patches. Users should also exercise caution when downloading software or opening e-mail attachments and avoid visiting suspicious or untrusted websites.

Trojans. A Trojan malware program is malicious software designed to appear harmless or useful, while containing malicious code that can harm the user's computer or steal sensitive data. The name *Trojan* comes from the story of the Trojan Horse in Greek mythology,[80]

---

80. The Trojan Horse, a seminal stratagem in Greek mythology, symbolizes deceit and ingenuity in warfare. According to legend, during the Trojan War, Greeks used a large wooden horse, ostensibly left as a peace offering, to stealthily infiltrate the city of Troy. Concealed within its hollow body were armed soldiers who, under the cover of night, emerged to open the city gates, allowing the Greek forces to enter and ultimately conquer Troy. This tale, epitomizing the use of subterfuge and surprise in overcoming an adversary, has since become a metaphor for any trick or strategy that causes a target to invite a foe into a securely protected bastion or space.

in which soldiers were hidden inside a wooden horse and then secretly entered a fortified city to carry out an attack.

Similarly, a Trojan can enter a user's computer system disguised as a legitimate program or file, often through social engineering tactics such as phishing e-mails or fake downloads. Once installed, the Trojan can perform various malicious actions, such as stealing passwords or personal information, creating backdoors for remote access by hackers, or deleting files. Unlike viruses or worms, Trojans do not self-replicate and must be manually installed by the user or through some other means of social engineering.

Advanced Persistent Threat. An Advanced Persistent Threat (APT) is a targeted cyber-attack carried by skilled and well-funded adversaries, such as state-sponsored actors or organized criminal groups. APT attacks are characterized by their use of sophisticated techniques, including manipulating people to reveal confidential information or do specific acts, zero-day exploits, and custom malware to gain continued access to a targeted network or system over an extended period.[81]

APTs often involve a multi-stage attack process, starting with reconnaissance to identify vulnerabilities in the target's network or systems. The attacker then uses operatives using pretexts to gain initial access, typically by sending targeted phishing e-mails or by exploiting software vulnerabilities. Once inside the network, the attacker installs custom malware or backdoors to maintain continued access and to gather sensitive data, often

---

81. Ping Chen, Lieven Desmet, and Christophe Huygens. "A Study on Advanced Persistent Threats," in proceeding of the *15th IFIP International Conference on Communications and Multimedia Security*, September 2014, Aveiro, Portugal, pp. 63–72.

stealing intellectual property or other valuable information.

APTs are typically motivated by espionage or financial gain and can cause significant damage to targeted organizations. They are often difficult to detect and mitigate because they are specifically designed to evade traditional security measures and remain hidden within a network or system for long periods.

To defend against APTs, organizations typically implement a layered defense strategy that includes threat detection and response capabilities, regular security awareness training for employees, and continuous monitoring of network traffic and system logs.

Cyber security practitioners posit that the lack of awareness companies have of their exposure to intrusions facilitates spies using cyber espionage as a route. The pace at which new hacking technologies are being created and the general increase in companies' online presence is acknowledged as enablers.[82]

---

82. PriceWaterhouseCoopers, *Study on the Scale and Impact of Industrial Espionage and Theft of Trade Secrets through Cyber*, 2019.

— CHAPTER FIVE —

Prevention and Detection

B efore we discuss the specifics of preventing and detecting industrial espionage, it is worth reviewing the benefits of crime prevention in general. Crime prevention is a critical concern for businesses of all sizes regardless of commercial sector or industry.

Arguably, the primary reason crime prevention is essential for businesses is to prevent financial loss.[83] Crime can result in direct loss from theft, vandalism, fraud, and indirect losses, such as increased insurance premiums and lost productivity. These setbacks can devastate businesses, particularly, small businesses that may not have the financial resources to absorb the impact.

Crime can also damage a business's reputation. News of criminal activity can spread quickly, jeopardizing customers' trust in the business. In turn, this can impact the business's ability to attract and retain customers.

Moreover, crime can impact employee safety, decreasing morale and reducing productivity. Employees who do not feel safe in their workplace are unlikely to perform at their best and are liable to leave. We all know that replacing staff adds to increased costs. But by preventing crime, businesses can mitigate the negative impacts of criminal activity.

---

83. Timothy D. Crowe, *Crime Prevention Through Environmental Design, Second Edition* (Boston: Butterworth-Heinemann, 2000).

Awareness Programs

Returning our thoughts to industrial espionage, one of the first things that should be considered is a program for employee awareness.[84]

An industrial espionage awareness program aims to educate employees about the risks of espionage and how to identify and prevent it. Employees need to understand what constitutes confidential information and why it is essential to protect it. They should also be made aware of the methods used by industrial spies, which we discussed in the previous chapter.

Ideally, the training program should be mandatory for *all* employees, including senior executives, managers, and staff across various functions. The training can be provided through workshops, webinars, e-learning modules, or other interactive platforms. To be effective, the program should include practical exercises demonstrating the various scenarios likely to be encountered, including examples of real-world situations highlighting the risks and consequences of industrial espionage.

Like all specialist training programs, this type of program should be developed and delivered by experts in the field of industrial espionage. Some topics for an employee awareness program could comprise the following themes:

1. Confidential information—Employees should be able to identify what constitutes confidential information and how to mark, handle, and store it securely.

---

84. Kevin D. Mitnick and William L. Simon, *The Art of Deception* (Indianapolis, IN: Wiley Publishing, 2002).

2. Pretexts—Employees should be able to recognize and prevent attacks by people using a pretext, such as phishing e-mails, telephone calls, and text messages.[85]

3. Physical security—Employees should understand the importance of physical security, such as locking doors, securing documents, and protecting their computer and network access credentials.

4. Cybersecurity—Employees should know cybersecurity best practices, such as using strong passwords, enabling two-factor authentication, and avoiding public Wi-Fi networks.

5. Reporting—Employees should know how and to whom to report any suspicious activity, such as unauthorized access to company information or theft of physical or digital assets (e.g., laptop computers, tablets, cell phones, and USB drives).

6. Security Team Training—One would think an IT security team is at the pinnacle of skill and ability. But, like any practitioner, some are head and shoulders above others. The OPM penetration is one example where a slow-reacting security team was exploited. Training needs to be tailored to these employees, too. Evidence of these firms' ongoing IT staff training should be verified if contractors are used.

Experience in fields as diverse as the military, government, education, and, of course, businesses have shown that by implementing an employee security awareness program, these organizations can experience

---

85. The use of pretexts is, arguably, one of the most successful ways of penetrating a company's security defenses. Greg Hauser, *The Pretext Manual* (Austin, TX: Thomas Investigative Publications, 1994).

improved outcomes. Nevertheless, conducting such a training course is not an end; such programs should be reviewed regularly and updated to reflect changes in the threat landscape.

Document Classification

It crucial to have a document security classification system that protects sensitive information from unauthorized access, disclosure, or misuse. A document security classification system is a way of labelling and categorizing hard-copy documents and electronic materials based on their level of confidentiality.[86]

Hard copy documents, such as contracts, financial reports, and personnel files, can be easily misplaced or stolen, compromising the security of their information. Without a document security classification system, employees will find it difficult to ascertain the document's level of confidentiality and proper handling procedures for that level of sensitivity. This can result in the careless handling, storage, and disposal of sensitive documents. In some jurisdictions, the law requires businesses to take reasonable steps to protect their intellectual property to be able to assert their legal rights over the material (i.e., in the case of theft). This relies on a classification system.

A document security classification system assigns classification levels to each document, such as top secret, secret, confidential, or unclassified. The system also provides guidelines for handling, storing, and disposing each classification level.[87] For example, a top-secret

---

86. Hank Prunckun, *Counterintelligence Theory and Practice, Second Edition* (Lanham, MD: Rowman & Littlefield, 2019).

87. Henry W. Prunckun, *Information Security: A Practical Handbook on Business Counterintelligence*, 2019.

document must be stored in a secure location with limited access. In contrast, a confidential document may only be shared with individuals with a *need-to-know*.

Electronic materials, such as e-mails, spreadsheets, and databases, pose a different security challenge. They can be easily accessed, copied, and distributed, making it difficult to keep sensitive information confidential.

Need-to-Know Principle

The need-to-know principle is a fundamental concept in information security that limits access to information only to those who require it to perform their jobs. The processes and procedures help minimize the risk of unauthorized access, data breaches, and other security incidents.[88]

The need-to-know principle is applied across various industries and contexts, from government agencies and military organizations to private businesses and non-profit organizations. In the military, for example, the need-to-know principle is a critical component of operational security (abbreviated OPSEC). This helps to prevent sensitive information from falling into the hands of the enemy and potentially compromising the safety of military personnel.

In business, the need-to-know principle protects financial data, trade secrets, and proprietary information. It is also an important component of regulatory compliance in many industries. For example, healthcare organizations must comply with government legislation

---

88. Hank Prunckun, *Counterintelligence Theory and Practice, Second Edition*, 2019.

that requires them to implement appropriate safeguards to protect patient information.

Organizations typically use various security measures, such as access controls, authentication mechanisms, and encryption, to implement the need-to-know principle. Access controls, such as role-based access control (RBAC) and attribute-based access control (ABAC), can restrict sensitive information access to only those requiring it.[89]

Authentication mechanisms, such as username and password authentication and multi-factor authentication, can be used to verify the identity of individuals attempting to access sensitive information. Encryption can protect sensitive information in transit or storage, ensuring it remains confidential.

Data Encryption

Encryption is the process of encoding information so that it can only be read by those with the necessary key or password to decode it (i.e., access control). It is used ubiquitously to protect sensitive information.

One of the most common uses of this digital technology is to secure online transactions—not only online banking and shopping but also the work employees perform when working remotely. It works in online transactions by using complex algorithms to scramble the information sent over the Internet. Without encryption, this information could be intercepted by hackers, who could then use it for fraud. However, encryption protocols such as Secure Sockets Layer (SSL) and Transport Layer

---

89. Vincent Hu, *Attribute-Based Access Control* (Boston: Artech House, 2017).

Security (TLS) ensure this information is encrypted during transmission,[90] making it extremely difficult for hackers to access.

Another encryption application is securing personal data, such as passwords. Many websites use encryption to store this information securely, often using one-way hashing algorithms that convert the original data into an unreadable format that cannot be reversed. This ensures that even if a database is breached, the sensitive information will not be usable by the attackers.

Encryption is also used in e-mail communication to protect messages from unauthorized access. By using encryption, the contents of an e-mail are transformed into an unreadable format that can only be deciphered by the intended recipient, who has the necessary decryption key. This is particularly important for businesses and governments that exchange sensitive information via e-mail.

Finally, encryption is commonly used in cloud computing, where data is stored on remote servers owned by third-party providers. By encrypting the data before it is uploaded to the cloud, users can ensure that their sensitive information remains protected even if unauthorized individuals penetrate the cloud service.

Firewalls

Firewalls can be either software applications or hardware devices used for access control. They act as a barrier

---

90. SSL and TLS are cryptographic protocols that establish a secure and encrypted connection between a client and server. They ensure that data transmitted over the Internet remains private by protecting it from eavesdropping or tampering. TLS is the successor to SSL, though both were in wide use at this writing.

between a company's internal network and the Internet by monitoring and controlling traffic across the network to prevent unauthorized access.

Firewalls can block traffic from specific IP addresses, ports, or protocols and can even inspect the contents of data packets for suspicious activity. In this regard, firewalls provide an additional layer of security for remote employees, who can access company resources outside the office.

Authentication Techniques

By the nature of remote working, as well as accessing data held centrally (when working from a business's premises), the issue that presents information security is how to allow access safely. In the analogue days, a person had to present at the "records department" with some authority.

However, with the omnipresence of centralized computing, identifying an authorized person first relied on issuing a user ID and password. When users failed to select or change a strong password at intervals, security administrators forced them to write these requirements into the application's login code.

Yet, with every defensive strategy, an industrial spy can analyze the situation to develop a way around it. To negotiate a way around passwords came programs that attack the logon application with hundreds of passwords per second, known as a *brute force attack*.[91] These "guesses" can be dictionary-based or based on some sequence of word/number/symbol combinations. Given the number of these programs available for download on

---

91. Lalit Kumar and Neelendra Badal, *Minimizing the Effect of Brute Force Attack* (London: Lambert Academic Publishing, 2021).

the Internet, it would suggest that this method continues to be successful.

Nevertheless, phishing is arguably the most effective technique. It is hard to believe that anyone with a smart device or a computer has not been subjected to a phishing attack; the number of employees still take the bait is numerous.

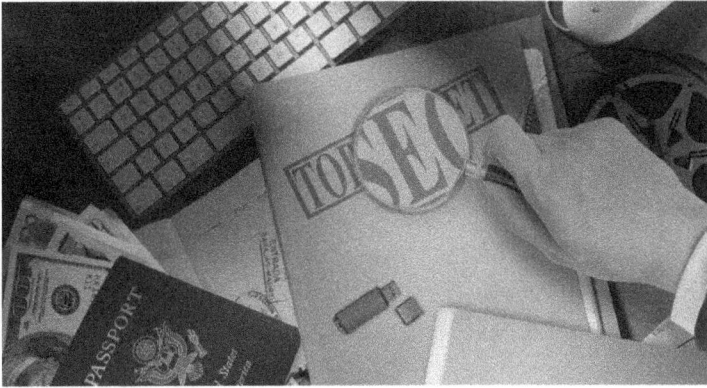

Figure 1—Personnel records are a very attractive target. Courtesy of the FBI.

For a business, it just takes one careless employee. Take the case of the U.S. Office of Personnel Management (OPM) hack that occurred between December 2014 and April 2015. It resulted in the theft of sensitive information belonging to millions of current and former U.S. government employees.[92]

It was a very attractive target because the OPM is the federal government agency responsible for managing the

---

92. Neil Daswani and Moudy Elbayadi, *Big Breaches: Cybersecurity Lessons for Everyone* (New York: Apress Media, 2021).

civilian workforce, conducting security background checks, and maintaining personnel records.

Evidence pointed to Chinese state-sponsored hackers who stole a range of data, including Social Security numbers, personal financial information, and security clearance data. The breach affected over 21.5 million people, including contractors and their families.

The stolen data could be used for identity theft and other criminal activities, and it would also have been a prized acquisition for espionage and other intelligence operations.

How can these data be used for future intelligence ops? Consider *spear-phishing*. Unlike phishing, a "broadcast" approach where hundreds of e-mails containing malicious code are sent, spear-phishing is a targeted e-mail. It is sent to a specific person, so background information makes the spear-fishing e-mail more convincing and more likely to succeed in penetration.[93]

It was concluded that an OPM contractor—KeyPoint—had one of its employee's logon access compromised, which, in turn, allowed access to OPM's network. Without going into the details, suffice it to say that if a more secure user authentication had been used, it would be highly unlikely to have occurred.

Such systems are known as two-factor authentication. Compare: one-factor authentication is a unique word or phrase that a user-employee knows; two-factor

---

93. The OPM penetration could be classified as a reconnaissance operation. Such information is generally used to formulate contextually relevant spear-phishing emails that contain a malware payload. The malware installs itself on the target's network/server and allows the industrial spy to launch its attack-in-chief.

authentication comprises the one-factor step, plus it requires the employee to possess something that verifies the first challenge, like a company ID card containing a security code or a security fob that produces a random code every, say forty-five seconds. Three-factor authentication builds on the two-factor techniques by adding a direct, physical connection to the employee, such as a fingerprint, voice print, or retina print.

Detection Technology

Data loss prevention (DLP) software is a suite of computer applications designed to protect sensitive data from unauthorized access. DLP software works by monitoring data usage, identifying potential security threats, and taking steps to prevent data loss.[94]

The first step in developing a business's DLP software is identifying sensitive data. This gets back to our discussion on classifying information. Once identified, DLP software monitors how this data is used, who has access to it, and where it is stored.

DLP software uses various techniques to monitor data usage by monitoring access, transfers, and replication patterns. For instance, DLP software can monitor network traffic to identify when data is transferred outside the organization, especially via e-mail or file-sharing services.

DLP software can also identify when unauthorized users are accessing sensitive data or when data is being used unusually. This includes identifying when data is being accessed outside of regular business hours when a

---

94. The Art of Service, *Data Loss Prevention: A Complete Guide—2021 Edition* (Brendale, Queensland: The Art of Service, 2021).

large amount of data is accessed, or when data is accessed from an unusual location.

Once potential security threats are identified, DLP software can take steps to prevent data loss. This includes blocking access to the information, alerting security personnel, and automatically encrypting or quarantining data.

A key feature of DLP software is its ability to enforce security policies. This includes policies around data access, storage, and usage. For example, DLP software can enforce policies requiring users to use secure passwords, encrypt sensitive data, and limit data access per the business's need-to-know principle.

Non-Disclosure Agreements

It is widely known that competitors hire a business's employees. Competitors do this to exploit the employee's knowledge acquired while in the rival's employ. It follows that employees with specific information, or those with years of experience, are targeted.

By way of example, take the case of Chinese firms that initiated a U.S.-based recruitment program (including the practice known as "headhunting") to attract personnel with knowledge of the sought-after intellectual property. The FBI reports the Chinese to have established what appeared to be legitimate recruiting firms that then placed Chinese "operatives" into targeted businesses to steal their IP.

These practices call for a business's lawyers to draft a non-disclosure agreement.[95] A non-disclosure agreement

---

95. Paul Witman, "The Art and Science of Non-Disclosure, Agreements," in *Communications of the Association for*

is a legal contract between an employer and an employee. It outlines the confidential information the employer will share for the employee to do their job, but it must remain private. These agreements establish how long the IP needs to be kept secret and the consequences for breaching the contract.

By signing a non-disclosure agreement, the employee agrees to keep the confidential and not disclose it to third parties. These agreements are legally binding contracts that are enforceable in civil courts should the employee breach the agreement.

International Agreements and Treaties

One of the most notable international agreements regarding industrial espionage is the *Economic Espionage Act, 1996* (EEA). The EEA is a U.S. federal law that criminalizes the theft of trade secrets, allowing the U.S. government to prosecute individuals and companies for breaches.

According to the EEA, the maximum penalty for individuals guilty of industrial espionage is fifteen years in prison and a fine of up to US$5 million. For corporations, the fine can go up to US$10 million. A view among criminologists is that the EEA has successfully reduced the incidence of industrial espionage in the United States and has served as a model for other countries to follow.

The Convention on the Protection of Trade Secrets is an international agreement on industrial espionage, also known as the "Paris Convention." The Paris Convention

Information Systems, Volume 16, Article 11, August 2005, pp. 260–269.

was signed in 1883 and has been updated several times. The Convention is a multilateral agreement that protects intellectual property rights, including trade secrets. It provides a framework for protecting sensitive trade information by setting minimum standards that signatory countries must meet. As of 2021, 177 countries have signed the Paris Convention.

Further, the World Trade Organization (WTO) has rules that address industrial espionage. The *Agreement on Trade-Related Aspects of Intellectual Property Rights* (the TRIPS Agreement) is one such pact.[96] The TRIPS Agreement is a multilateral treaty that requires member countries to have laws and regulations in place to protect trade secrets and to provide remedies for their unauthorized use. As of 2022, some ninety-five countries have signed the TRIPS Agreement.

In addition to these agreements, some countries have established bilateral arrangements to address industrial espionage. For example, in 2015, the United States and China signed a bilateral agreement. The agreement provides a framework for cooperation between the two countries and sets out the obligations of the two countries to protect each other's trade secrets.

Finally, the United Nations (UN) has established a program called the *United Nations Conference on Trade and Development* (UNCTAD), which provides technical assistance to developing countries on issues related to intellectual property rights. The UNCTAD conducts

---

96. World Trade Organisation, *Agreement on Trade-Related Aspects of Intellectual Property Rights,* as amended (Geneva: World Trade Organisation, January 23, 2017).

research and analysis on the impact of industrial espionage on developing countries.

Legal Recourse

As we saw, the U.S. *Economic Espionage Act* makes industrial espionage a federal crime. Still, in addition to criminal penalties, the EEA provides civil remedies, such as injunctive relief and monetary damages. Victim companies can file a lawsuit under the EEA to recover damages for losses suffered because of the theft of their trade secrets.

Australia has similar laws in place to protect businesses from industrial espionage. The Australian Federal Police[97] (AFP) investigate allegations of espionage, and those found guilty can face up to ten years in prison. In addition to criminal penalties, companies can pursue civil remedies through the Australian legal system. The laws in Australia allow for the recovery of damages suffered.

Industrial espionage is a criminal offence under the *Computer Misuse Act, 1990* in the United Kingdom. The act provides penalties of up to two years imprisonment and/or a fine. The U.K. also has a civil legal system that allows companies to sue for damages resulting from

---

97. The Australian Federal Police serves as the principal federal law enforcement agency of Australia, established in 1979 through the amalgamation of the Commonwealth Police and the Australian Capital Territory Police. Its responsibilities encompass investigating and preventing crimes against the Commonwealth, including terrorism, drug trafficking, human trafficking, cybercrime, and financial crimes. Operating both within the country and internationally, the AFP plays a crucial role in national security, providing policing services to Australian territories and contributing to global efforts against transnational criminal activities.

industrial espionage. To succeed in a civil action, a company must show that its intellectual property has been misused and suffered losses as a result.

Canada has laws to protect businesses from industrial espionage, including the *Security of Information Act and the Criminal Code of Canada*. These laws provide criminal penalties for those who steal trade secrets or engage in other forms of industrial espionage. In addition to criminal penalties, Canadian businesses can pursue civil remedies for damages suffered by industrial espionage.

New Zealand has laws in place to protect businesses from industrial espionage.[98] The New Zealand government takes a strong stance on industrial espionage, and those found guilty can face up to seven years in prison. In addition to criminal penalties, New Zealand businesses can pursue civil remedies through the legal system.

In all the countries mentioned, businesses must be able to prove that their intellectual property has been stolen or misused to pursue legal action successfully. This can be challenging because industrial espionage is a covert, difficult to detect. However, with legal representation and evidence, companies can recover damages while protecting their intellectual property from future theft.

---

98. The New Zealand Police, as a national law enforcement body, primarily focuses on maintaining public order, preventing, investigating, and prosecuting crime within New Zealand. While there is no specific public record of the New Zealand Police's involvement in industrial espionage cases, their role could potentially extend to investigating and addressing such incidents if they constitute criminal activities under New Zealand law.

— CHAPTER SIX —

Legal and Ethical Implications

B ecause industrial espionage involves obtaining confidential information without authorization; these acts raise legal and ethical issues. We will survey some of these critical matters here.

Patents, Trademarks, and Copyrights

Laws, such as patents, trademarks, and copyrights, protect intellectual property. Patents, trademarks, and copyrights are legal mechanisms that protect different types of intellectual property.

Patents protect inventions like machines, processes, chemical compositions, and pharmaceuticals. A patent gives the business inventor the right to prevent others from making, using, selling, or importing the invention for a limited time, usually twenty years from the filing date. In exchange for what is essentially a monopoly, the business must disclose the invention in detail in its application so that others can learn from it—or steal it...[99]

Trademarks protect names, logos, and slogans that are used to identify goods or services from others in the market. A trademark provides the right to prevent others from using a similar mark in connection with similar goods or services. A trademark can last indefinitely as

---

99. Consequently, some businesses decide not to file for a patent, opting to keep their invention a trade secret.

long as it is in use and the owner continues to enforce their rights.

Copyright protects original works of authorship, such as books, music, software, and art. A copyright allows the creator to prevent others from reproducing, distributing, performing, or displaying the work without permission. Copyright protection lasts for the author's life plus a certain number of years after death, typically seventy years in many countries.

Overall, these legal mechanisms help promote innovation and creativity by incentivizing inventors, artists, and entrepreneurs to invest time and resources into developing new ideas and products.

While these forms of protection can help prevent industrial theft, they may not always be sufficient. An industrial spy may still try to infringe these rights. In such cases, the owner can take legal action, which seeks damages for any losses suffered due to the infringement.[100]

This protection, in theory, is reassuring, but when it comes to state actors (or state-sponsored actors), these laws may be impossible to enforce. Take the cases of China, North Korea, Iran, and Russia. Although these countries have their legal framework for protecting intellectual property, there are concerns across the international legal community about the effectiveness of their enforcement actions.

Although some think China has made some progress in improving its intellectual property laws and enforcement

---

100. Gillian Dempsey, *Industrial Espionage: Criminal or Civil Remedies* (Canberra, ACT: Australian Institute of Criminology, 1999).

mechanisms, concerns remain about counterfeiting and piracy in that country, particularly regarding products like software and luxury goods.

North Korea is universally known for its isolationist policies and lack of transparency;[101] this makes it difficult to say with certainty what its policies are regarding intellectual property. However, there have been reports of North Korean entities perpetrating cyber penetrations to steal intellectual property.

Iran has a legal framework for protecting intellectual property, but international opinion is that enforcement is weak in practice. Piracy and counterfeiting have been reported to be widespread throughout the country, particularly regarding copyrighted material like music and movies.

Russia is viewed as a violator of intellectual property laws, with piracy and counterfeiting widespread nationwide. There have also been concerns about Russia's lack of enforcement mechanisms and its government's unwillingness to apply controls.

It would not be unreasonable to say that these countries hold a different ethical perspective on intellectual property rights, especially when ignoring these rights will help promote innovation and economic growth in their own country. This is especially true when it comes to disregarding the rights of businesses based in Western liberal democracies.

---

101. North Korea is widely recognized for its isolationist stance and opacity in governance. This approach limits external influence and information flow, contributing to its enigmatic presence on the global stage.

Violation of Trade Secrets

When a trade secret is misappropriated, the affected business can seek legal remedies through civil courts. The plaintiff must show several "elements" of the offense to establish a claim.[102]

The chief element is demonstrating that the information was a trade secret, meaning it was valuable, not generally known or readily ascertainable, and subject to reasonable efforts to maintain its secrecy. (On this last point, see our previous discussion about *security intelligence.*)

The next element is misappropriation. The plaintiff must show that the respondent acquired, disclosed, or misused the trade secret or breached a confidential relationship (e.g., in contravention to a non-disclosure agreement or employment contract).

The final element is that of damage or harm. It needs to be proven that the misappropriation caused harm to the plaintiff's business or the value of the trade secret.

If a plaintiff can establish these elements, they may be entitled to relief, such as an injunction preventing further use or disclosure of the trade secret, damages for lost profits or other harm, and court costs and legal fees.

Breach of Confidentiality

Many businesses have agreements to protect confidential information, such as non-disclosure agreements (NDA). An NDA is a legal contract between parties that outlines information they want to keep private.

---

102. The specific legal requirements for taking a civil action for the misappropriation of trade secret will vary by jurisdiction.

NDAs can be one-way, where only one party shares confidential information, or mutual, where both parties share confidential information. In this regard, the agreement identifies the confidential information, the time the agreement will be in effect, and the consequences for breaching the agreement.

Because NDAs are legally binding contracts, they can be enforced through civil legal action if a party breaches the agreement.

Ethical Concerns

Industrial spying raises ethical questions about the role of competition in business, the responsibilities of companies to protect their proprietary information, and the legal and social frameworks that govern these practices.[103]

In many countries, industrial espionage is illegal and punishable by law. However, enforcement can be difficult, and some argue that the penalties are not severe enough to deter the practice—those who do the spying have said that they see a fine as one of the costs of doing business.

Some observers argue that stealing confidential information is simply part of the game of competition. In some industries, industrial espionage is considered a common practice, while in others, it ranges from being frowned upon to being classified as taboo.

With some industries' attitudes of acceptance, it is incumbent on businesses to protect their proprietary

---

103. John J. McGonagle, "Competitive Intelligence," in Peter C. Oleson, editor, *AFIO's Gudie to the Study of Intelligence* (Falls Church, VA: Association of Foreign Intelligence Officers, 2016), pp. 376–378.

information. Therefore, companies are responsible for taking reasonable measures to guard against penetration.[104]

International Laws and Treaties

Industrial espionage can affect international relations and strain diplomatic relations between countries. The victim country may see these actions as a violation of their sovereignty and an attack on their economy. This is because spying can harm the competitiveness of the victim country's economy, leading to retaliation through tariffs and other economic sanctions.

Diplomatic issues may require intervention from international bodies like the United Nations or the World Trade Organization. One of the more significant industrial spying treaties is the Agreement on Trade-Related Aspects of Intellectual Property Rights (TRIPS), adopted by the World Trade Organization (WTO) in 1994. TRIPS establishes minimum standards for protecting and enforcing intellectual property rights, including patents, trademarks, and copyrights. These standards help to prevent the misappropriation of trade secrets.

There are other international treaties, for example, the United Nations Convention against Transnational Organized Crime. This was adopted by General Assembly resolution 55/25 (November 15, 2000) and has become the leading international instrument in addressing transnational organized crime, which includes helping to prevent the theft of trade secrets.

---

104. Companies bear the responsibility of implementing adequate safeguards to prevent unauthorized access or breaches. This duty encompasses adopting robust security protocols and measures to protect sensitive information and infrastructure.

The Convention on Cybercrime (also known as the *Budapest Convention*), which was adopted by the Council of Europe in 2001, also includes anti-theft provisions relating to computer systems.

Lastly, many countries have laws related to industrial espionage. For example, in the United States, the *Economic Espionage Act, 1996* makes it a federal offense to steal or attempt to steal trade secrets for the benefit of a foreign government or company.

Spies who engage in industrial espionage can face severe consequences. In many countries, stealing confidential information can result in criminal charges, and if found guilty, the person(s) could be fined and/or imprisoned. This can have long-lasting consequences for an individual's career and personal life, such as loss of employment, reduced opportunities for advancement, and damage to their professional reputation.[105]

Here is such a case—the theft of trade secrets from Monsanto. We understand that a former company employee, Mo Hailong, was accused of stealing secrets related to Monsanto's genetically modified corn seeds. Hailong was a Chinese national and part of a group of people allegedly trying to steal proprietary information from several U.S. agriculture companies.

The theft was reported to have been discovered by Monsanto after the company noticed suspicious activity on their experimental fields in Iowa, where the genetically modified corn seeds were being tested. The company then launched an investigation.

---

105. This can significantly impact their future earning potential and career prospects.

The inquiry found evidence that Hailong and his accomplices had been taking photos of the corn, stealing seeds, and attempting to send them to China. Hailong was arrested in 2013 and pleaded guilty to conspiracy to commit economic espionage. In 2017, he was sentenced to three years in prison.

Impact of Industrial Espionage on Global Economics

Industrial espionage can have significant impacts on global economics in several ways. The first is the loss of intellectual property. Businesses invest significant time and money in research and development to develop new products, services, and processes. Loss of this ability harms innovation. Companies that invest in research and development cannot afford to lose their inventions if they are to continue to innovate and thrive.

Industrial espionage can damage a business's reputation. For instance, if a company is known to have been a victim of espionage, it may lose the trust of its customers, partners, and investors—share prices may fall—resulting in a loss of sales and profits.[106]

Industrial spying also has national security risks, especially if the stolen information is related to defense or critical national infrastructure. In some cases, foreign governments have been implicated, which has had geopolitical implications.

---

106. Christopher Burgess and Richard Power, *Secrets Stolen, Fortunes Lost: Preventing Intellectual Property Theft and Economic Espionage in the 21st Century* (Burlington, MA: Syngress Publishing, 2008).

— CHAPTER SEVEN —

Outlook

A s businesses become more global, it follows that competition will rise. As competition increases, so will the threat of industrial espionage. But as discussed earlier in this book, industrial espionage is not new. Throughout history, businesses have sought to gain an edge over their rivals. However, in today's digital age, the stakes are higher. With the rise of technology, it has become easier for malicious actors to access sensitive information. This has led to an increase in the frequency of attempts to penetrate businesses' defences.[107]

Intellectual property is undoubtedly the lifeblood of businesses, and the theft of proprietary information can be devastating and impact national security. And in this regard, there are many news reports where foreign governments or state-sponsored entities carry out industrial espionage to advance their strategic goals.

Is it a hopeless situation? No, certainly not. The state of spying underscores the importance that businesses take steps to protect themselves. Even simple crime prevention measures such as implementing strong cybersecurity measures, conducting background checks on employees and contractors, and developing protocols for handling

---

107. I.I. Androulidakis and Fragkiskos–Emmanouil Kioupakis, *Industrial Espionage and Technical Surveillance Counter Measurers* (London: Springer, 2016).

sensitive information will go a long way to prevent loss. In addition, being aware of the signs of industrial espionage—unexplained data breaches or unusual activity on company networks—will signal that the effort to attempt penetration isn't worth the risk. By remaining vigilant, businesses can reduce their probability of falling victim to industrial espionage.[108]

Nonetheless, preventing industrial espionage is not always easy. Malicious actors constantly develop new tactics, and the line between legal competitive intelligence and illegal industrial spying can be blurry. This makes distinguishing between legitimate competitor intelligence business practices and unlawful activity difficult.

For the spy-for-hire, the espionage landscape has changed dramatically over the years. Espionage has evolved from the old days of using "dead drops" and arranging clandestine meetings to the current era of digital surveillance and cyber-attacks.

So, what does the future hold for the future spy? The task of gathering intelligence to provide strategic advantage to clients will remain. Another sure thing is that the demand for their services will continue to grow. In a world where information is power, businesses and governments always need someone to collect secret intelligence.

However, the methods and techniques industrial spies use will undoubtedly change. In the past, espionage was often a game of "cat-and-mouse," with agents trying to outmaneuver each other in the physical world.

---

108. Ronald L. Mendell, *The Quiet Threat: Fighting Industrial Espionage in America, Second Edition* (Springfield, IL: Charles C. Thomas, Publisher, 2011).

Nevertheless, with the advent of new technologies, the "battlefield" of espionage has shifted, in large part, to the digital realm. Today, skilled hackers can gain access to sensitive information without ever leaving their desks, and indications are that this trend will continue.

Rise of Private Intelligence Firms

Legitimate private intelligence firms are an alternative for businesses seeking a competitive edge. Also known as corporate intelligence or business intelligence firms, these inquiry agents have become increasingly popular.

These firms provide research and analysis services to private clients such as corporations, law firms, and wealthy individuals. The rise of these firms can be attributed to a few prime factors—the increasing complexity of global business and the proliferation of data.

Take as an example the case of the private intelligence firm Stratfor. It is a U.S.-based company founded in 1996 and provides intelligence on worldwide political, economic, and security issues. Stratfor's clients include Fortune 500 companies, government agencies, and military organizations. It was once described as "The Shadow CIA."[109]

Another example is Kroll, a New York-based research firm founded in 1972. It, too, provides investigative, intelligence, and risk management services to clients in various industries.

---

109. Jonathan R. Laing, "The Shadow CIA," October 15, 2001. Available at: https://www.barrons.com/articles/SB100292755743 4087960. Accessed February 21, 2023.

Legitimate private intelligence firms have gained attention due to their involvement in high-profile investigations. In 2011 it was revealed that the private intelligence firm Palantir Technologies had provided services to the U.S. government in tracking Osama bin Laden. Palantir's software was used to analyze large amounts of data to identify potential leads for searchers of the notorious al-Qaeda leader, Osama bin Laden.[110]

Nonetheless, private intelligence firms have faced criticism for their lack of transparency and potential conflicts of interest. Some critics have argued that firms can engage in unethical or illegal activities to obtain information. At the same time, other observers have raised concerns about the potential for private intelligence firms to influence public policy. A standout example is the British research firm Cambridge Analytica.

Cambridge Analytica was a political consulting firm accused of improperly harvesting data from millions of *Facebook* users without consent to influence Donald Trump's 2016 U.S. presidential election campaign. The controversy began when a whistle-blower, Christopher Wylie, revealed that the firm had obtained data from an app called *This Is Your Digital Life*, created by a researcher named Aleksandr Kogan.[111]

---

110. Mathieu Rosemain, "France Seeks Own Alternative to Palantir Data Firm in Helping Fight Terrorism," Reuters *Aerospace and Defense*, November 28, 2019. Available at: https://www.reuters. com/article/us-france-palantir-surveillance-idUSKBN1Y11NI. Accessed March 27, 2013.

111. Carole Cadwalladr and Emma Graham-Harrison, "Revealed: 50 million Facebook Profiles Harvested for Cambridge Analytica in Major Data Breach," *The Guardian*. Available at: https://web.archive.org/web/20201109034648mp_/https://www.t

Kogan had obtained the data through *Facebook*'s API, which at the time allowed third-party apps to access user data and the data of their friends. Kogan is said to have sold these data to Cambridge Analytica, which used it to create psychological profiles of voters and target them with political ads.

The controversy led to investigations by governments and regulatory bodies in the U.S. and U.K., resulting in *Facebook* facing intense scrutiny over its data privacy practices. Cambridge Analytica was later shut down, and its parent company, SCL Group,[112] filed for bankruptcy.

The scandal increased public awareness of the importance of data privacy and the need for stronger regulation of social media platforms. So, while there are isolated cases where these firms have misstepped, ethical firms provide valuable insights for their clients.

Artificial Intelligence

Another trend likely to shape espionage's future is the increasing use of artificial intelligence (AI). These technologies are now used in the ethical realm of competitive intelligence to sift through vast amounts of data, looking for patterns to reduce uncertainty for future events. As these technologies become more sophisticated, they will become more valuable and give our discussion about Cambridge Analytica to the industrial spy.

For industrial spies, AI can mean they can provide the most accurate and helpful intelligence. Granted, they will

---

heguardian.com/news/2018/mar/17/cambridge-analytica-face book-influence-us-election Accessed March 27, 2023.

112. Strategic Communications Laboratories (SCL) had described itself as a global election management agency.

need to be able to navigate complex datasets and identify the most important information. Still, as Cambridge Analytica showed the world, some data scientists can do this for them. For nation-states and spies sponsored by these states, AI can break down another nation's cyber defenses.[113]

Ethical Dilemmas

As the world of espionage becomes more intricate because of technology, we will likely see an increasing number of ethical dilemmas arise. This will not be an issue for industrial spies who ignore such considerations, but moral dilemmas will impact those who practice business intelligence—the legal form of competitive intelligence. Practitioners in the latter category must navigate these murky waters carefully, ensuring they do not cross the line that could damage their or clients' reputations.

Defensive Practices

As the threat of industrial espionage continues to evolve, we likely expect to see increased efforts to combat the practice. This may include more substantial legislated penalties for those who engage in industrial espionage, increased regulation of specific industries, and increased collaboration between government security and intelligence agencies and the private sector to share attack information.

Sharing attack information helps stem what sometimes appears to be an avalanche of malware sent through e-mail services. These e-mails carry malware payloads that attempt to steal data, install spyware, sabotage computer

---

113. Hedieh Nasheri, *Economic Espionage, and Industrial Spying* (Cambridge, UK: Cambridge University Press, 2005).

systems, or lay in wait to undertake some nefarious future task.[114]

Moreover, businesses will be rethinking their approach to intellectual property. In the past, many businesses focused solely on securing their intellectual property through patents and other legal instruments. Today, this is not enough. Businesses will likely take a proactive approach by investing in training awareness programs and cybersecurity and developing more robust internal document controls.

Regardless, lawyers still play an important part in defensive measures. What springs to mind immediately is providing advice (e.g., drafting contracts) about software leasing arrangements for cloud-based computing and storage. Such legal agreements must ensure that the business's information is stored on physically and electronically secure hardware and transferred over secure networks.

Other Key Developments

The Internet of Things (IoT) has been hailed as a leap forward in efficiency. Connecting so-called "smart devices," like office equipment, medical devices, refrigerators, televisions, cars, and so on, via the Internet can bring about unrealizable productivity. Data from these devices are transmitted without human action, sometimes without a business's knowledge.

The issue that played out with Cambridge Analytica has the potential to play out with the information found in the IoT. The interconnected nature of smart devices, with

---

114. Randolph Kahn, *Information Nation: Seven Keys to Information Management Compliance, Second Edition* (New York: Wiley Press, 2009).

limited security, makes trade information vulnerable to industrial spies.

"In determining the value of protected information, a company must look at the investment in the time and research that went into gathering and or creating the information. This includes capital spent, salaries of individuals who worked on the project, plus the cost of storing and securing the information. The company must also evaluate the replacement cost to the organization, if possible. The final area to consider is the monetary loss if the information were to be used by another organization."[115]

Information can have value It is axiomatic that the scarcity of a piece of information determines its worth and the more desirous it becomes. Enter the "dark web," where almost anything, including trade secrets, can be bought and sold.[116]

Given this freedom from scrutiny, it is often used for anonymous and illegal activities such as buying and selling drugs, firearms, or stolen information. Reports indicate that industrial spies can buy the software "tools" (i.e., computer programs) that allow them to spy on people

---

115. Daniel J. Benny, *Industrial Espionage: Developing a Counterespionage Program* (Boca Raton, FL: CRA Press, 2014), p. 7.
116. The *dark web* is a part of the Internet that is only accessible through special software. Information found on the dark web is not indexed by conventional search engines. While not all content on the dark web is illegal, it is a space where users can operate with a high degree of anonymity and without oversight, thus, making it a center for criminal activity.

and exfiltrate confidential corporate data from rival firms.[117]

Nation-states have even subcontracted spying tasks to hackers to maintain "distance" from their operation (i.e., deniability). This has been described as a "new frontier" where states contract industrial spying to privateers—cyber-space mercenaries.[118]

---

117. Nearchos Nearchou, *Combating Crime on the Dark Web* (Birmingham: Packt Publishing Ltd, 2023).

118. Tim Maurer, *Cyber Mercenaries: The State, Hackers, and Power* (Cambridge, UK: Cambridge University Press, 2018).

ABOUT THE AUTHOR

Dr Henry (Hank) Prunckun, BSc, MSocSc, MPhil, PhD, is an Adjunct Associate Research Professor of intelligence methodologies with the Australian Graduate School of Policing and Security, Charles Sturt University, Sydney. He is a former Australian government intelligence analyst and government-licensed private investigator. Dr Prunckun spent much of his twenty-eight-year operational career in tactical intelligence and strategic research. He also served operationally in physical and cyber security, white-collar crime investigation, and counterterrorism.

INDEX